GOATS

WRITTEN AND ILLUSTRATED BY

Wilfrid S. Bronson

SUNSTONE
PRESS

SANTA FE

Sunstone books may be purchased for educational, business, or sales promotional use.
For information please write: Special Markets Department, Sunstone Press,
P.O. Box 2321, Santa Fe, New Mexico 87504-2321.

Printed on acid free paper

Library of Congress Cataloging-in-Publication Data

Bronson, Wilfrid S. (Wilfrid Swancourt), 1894-1985.
  Goats / written and illustrated by Wilfrid Swancourt Bronson.
    p. cm.
  ISBN 978-0-86534-774-8 (softcover : alk. paper)
  1. Goats. 2. Goat breeds. I. Title.
  SF383.3.B76 2010
  636.3'9--dc22

                          2010033681

Published in
Santa Fe

WWW.SUNSTONEPRESS.COM
SUNSTONE PRESS / POST OFFICE BOX 2321 / SANTA FE, NM 87504-2321 /USA
(505) 988-4418 / ORDERS ONLY (800) 243-5644 / FAX (505) 988-1025

GOATS

"Goat." What do we think of when we hear that word? An animal, with a beard and horns, that butts? An animal that eats tin cans and clothes off the line and smells unpleasantly? That is the popular idea of a "billy goat." Some of it is right, part is wrong, and the rest need not be so.

A big billy, or buck goat, weighs over two hundred pounds and is the size of a Shetland pony. He is probably as proud of his beard as many a man. And to judge from the arch of his powerful neck and the tilt of his head, he wears his horns both as weapons and as a glorious crown. In wild nature he fights the other bucks to become the king of a flock of many does, or nanny goats, with all their half-grown kids and older goatlings. He loves this battling with the other bucks, for wild or tame, with horns or without, all goats delight to butt. Butting is as

7

much a part of being a goat as having a beard at one end and a short tail at the other.

So, from the beginning of their lives to the end, wherever more than one goat is found, bucks, does, and kids alike whack heads together, mostly in mock fights just for fun. But this play gives good practice to growing kids whose battles later on may sometimes be more serious, against rival goats or outlaw dogs, wolves, wildcats, and other goat-attacking enemies. Two grown-up does passing each other closely in the meadow appear to nod like ladies meeting on Main Street. But really, each lowers her head, if ever so slightly, to be ready to meet any sudden jab the other may give. If this comes when only one of them is in a butting mood, the other may leap straight up in the air (all four feet off the ground, light as an antelope) and come down running. Like as not though, there will be no jab as they pass on peacefully.

But often, especially when the weather is good and all the flock is feeling fine, two such friends pause and bump heads gently. They bump again less gently, then rather hard, and then harder. Finally they back away only to clash together again with terrific force. They rear up and come down slashing sideways, as though to tear each other

8

with the points of their horns, but nearly always meeting head-on in another shocking crash, as each matches the battering swipes of the other.

We wonder what can be left of the brains inside their jolted, ringing heads, for brains they surely have, and good ones. They are just as tame and affectionate toward kindly masters as are their relatives, the sheep. But goats on a farm remain, like their wild ancestors, of very independent mind. Far more than sheep, they like to follow their own ideas. Not only mischievous and playful but very intelligent, they are curious about almost everything and want to find out more. Thus, if not very carefully enclosed, some learn to untie knots and unlock latches to go exploring far from their pens and proper pastures. They will find ways to climb or jump out of, or over, or into, or onto, all sorts of places where they aren't supposed to be; your guest's fine car perhaps. They will sniff and taste things not meant for food, and sometimes eat them–the flowers in your neighbor's garden, for example.

9

The goat that seems to be eating tin cans is only finding out how the paper on them tastes. Of course, if it is a poor half-starved goat, forced to feed on rubbish and swill in some untidy slum, it may swallow quite a lot of paper just to pad its cramped insides, or eat fresh clothes from a wash line. Even a terribly hungry goat would rather eat a clean shirt than a dirty one.

Goats like variety, and on the farm as in nature, where there is plenty to choose from, they are finicky in their diet. Though they like apples very much, they may decline to taste one that you have bitten first. They will refuse food that has fallen on the barn floor, and always eat the fresher and sweeter leafy hay. They love cigarettes, to eat that is, as well as leaves of other flavors.

A RAISED PLATFORM WITH SEAT ATTACHED IS IDEAL — THOUGH A FEW PEOPLE ALWAYS DO THINGS THE HARD WAY.

MUSK DEER • GOAT MOTH • PECCARY • CIVET "CAT" • MUSK BEETLE • BEAVER • ALLIGATOR

Now as to flavors, the strong aroma that bucks have, though displeasing to people, is a charming scent to she-goats. The does themselves have no more odor than their own delicious milk. Only the he-goats have it, mostly in the breeding season, from early fall till nearly spring. But they aren't the only creatures that make themselves attractive with perfume to others of their kind. Certain deer and wild swine, beavers, civet cats, alligators, and various insects do, to name a few. It's a common thing in nature. Each kind has its own preferred aroma that it alone especially enjoys.

The odor of some bucks is hardly noticeable: others can be smelled for half a mile. Yet a goat breeder or dairyman can keep any buck smelling sweet enough by clip-

A VANDYKE GOATEE

ping his shaggy coat and trimming his bushy beard to a smart goatee, and then every few days washing him with a disinfectant in warm soapy water and spraying his shed or stall with a good deodorant. For an animal as important as a farmer's prize buck, the father of every kid born on the farm, this isn't too much trouble.

What about all the newborn kids? They generally come in pairs, a brother and sister. Let's just suppose each of a farmer's fifty she-goats has twins this year. That's fifty little does and fifty baby bucks. The baby does will grow up and give milk the farmer can sell to make a living. But for every fifty new does only one young buck will be needed. What's to be done with the other forty-nine? He can't raise them all. They would simply eat and smell him out of business.

Well, most buck kids will be kept awhile and fattened up for market. They are liable to reach people's dining tables labeled "lamb." A few may be raised to sell as pets. On each of these a veterinary doctor, or the farmer himself, performs a little operation to prevent it from ever having an unpleasant odor. It will never have any kid children either; but though growing up big and very strong, it will be both gentle and sweet-tempered. Such a buck is called a wether. Its affections—just like any dog's—will belong to the young folks that it plays with. It will make a perfect pet.

A full-grown pet wether can give its owner and the neighborhood children good fast rides in a goat cart or even carry a small child on its back. It will put up with endless mauling, for youngsters sometimes pull its tail or beard, not knowing any better. For hours on end it will not kick or threaten to butt. But let some unmannerly dog come barking at its heels, and it will stand on its hind legs, shaking its great head. The dogs usually are amazed and frightened, for when a big goat rears up, it is taller than a man. And when it comes down slashing with those mighty horns, the dogs see danger and keep clear. A wether goat can be a wonderful playmate, enjoying the fun as much as anyone else.

In parts of Africa, in Latin America, in the Far East and the Near East, and in Europe, people keep goats to supply themselves with milk and cheese. And more and more this is true in Canada and the United States, as our own goat dairying increases. Every year we have new goat farms and ranches producing milk, cheese, butter, ice cream, meat, leather, and fleece for weaving cloth.

Some farms, both here and abroad, are very big with many animals and hundreds of acres of good pasture. Their milk and cheeses compete with those produced from cow's milk in the markets of great cities. In other places, small flocks of goats provide most or all of the milk used by the people. Half a dozen nimble, nibbling goats can get about where a cow would break her neck and find food enough where she would starve. Thus, because of goats, even in rather barren areas there can be sweet and wholesome milk to drink or to make into cheese and butter. In many lands milk is drawn from other local animals, as shown in the picture. But goat's milk has special merit, which we shall notice farther on.

WATER BUFFALO

BOSSY

REINDEER

YAK

SHEEP

HORSE

CAMEL

MILK—
ANYBODY?

DONKEY

SPIKE-ANTLERED
DEER

IMPALLA
ANTELOPE

BUCK
GOAT

DOMESTIC
SHEEP
RAM

As you know, a herd of tame goats is made up mostly of milk-giving does. Most does are daintier, much more elegant creatures than their sires and sons. They remind us of deer somewhat, and even more of antelopes. Antelopes may be the distant cousins of goats and sheep, with a common ancestor in prehistoric times. Today there are various breeds of goats both wild and tame. And since all tame breeds descend from wild ones, let us look first at the wild kinds, along with the wild sheep and antelopes they most resemble.

As anyone can see (and hear), sheep and goats have much in common. There are enough differences though, so that usually we can say, "This is a sheep—that is a goat," when we view them side by side. Rams, the male sheep, have no beards, whereas the bucks of all but one or two breeds of true goats do. The horns of rams tend to coil round and round like heavy springs. Those of buck goats sweep up and out and back like curving oriental swords, some kinds also twisting like a corkscrew. Rams have no special odor, whereas he-goats decidedly do.

16

WILD SHEEP
BIGHORN
BHARAL
AOUDAD

But now consider the animals that look as much like one side of the family as the other. The Bharal or Blue sheep of Tibet in Asia and the Aoudad sheep of the Atlas Mountains in North Africa are two kinds that are very goatlike. They are related to the Bighorn sheep of our western mountains. Yet their horns, though heavy, don't coil as most sheep's do. Rather, they curve up, out, and backward like a goat's, though they don't twist either, as do so many goat's horns.

The Bharal is short-haired and wears no beard. The Aoudad has none either. But from his throat hangs a great ruff of beardlike hair all the way to his knees, as though he has put on false goat's whiskers that have slipped. Both of these wild sheep are often kept in zoos where you can see, close up, their very goatliness.

17

TUR

SPANISH
WILD
GOAT

PERSIAN
WILD
GOAT

So much for the goatlike sheep. Next come sheeplike
goats. The goat with the most "sheepish look" is a so-
called Tur of the Caucasus Mountains. Resembling the
Bharal sheep especially in his horns, the Tur seems to have
the problem whether "tur be or not tur be" a goat. But
his whiskers and perfumery decide the question of which
he is as far as we are concerned.

The Spanish wild goat seems free of the Tur's uncer-
tainty. Yet in his old age he may take on a Turish look
when the knobs on his horns wear smooth like a Tur's.

The Persian wild goat or Paseng is and looks entirely
goatish all his life. And all our varied domestic breeds,
except perhaps Angoras, are descendants of his race. Part
of the variety among tame goats comes from centuries of
man's selective breeding. But some of it is the result of
other kinds of wild goats breeding with the tame ones
over all the Persian goat's wide range. This goat is found

18

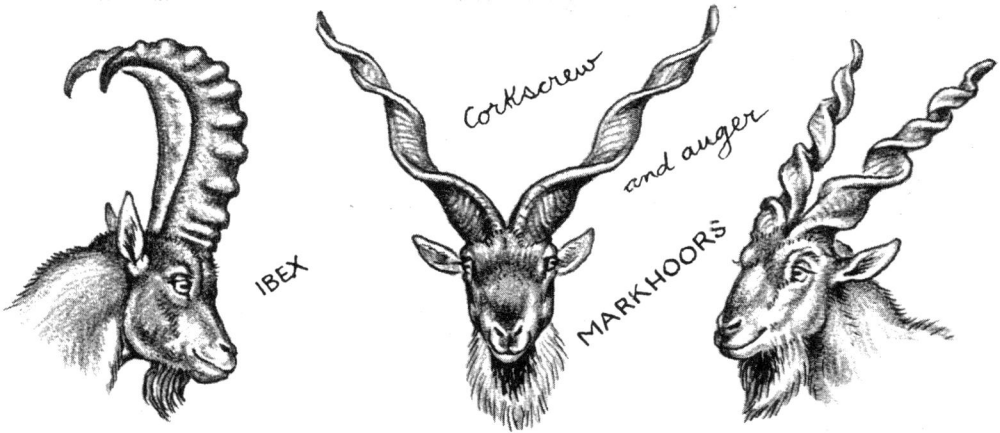

on mountains clear from Greece through Persia (now Iran), Afghanistan, and into India.

Two kinds of wild goats that sometimes mingle with tame Persian herds (especially in the breeding season) are the Ibex and the Markhoor. In either case, the tame king of the herd may often need the help of a goatherd to drive such visitors back to their higher mountains, for both Ibex and Markhoor, being wild, are extra tough and hardy, and their fighting equipment is superb. The stout horns of the Ibex rise very high while curving back. And all along the forward edge they bear thick ridges for rasping and breaking an opponent's weapons. They remind one of the war clubs carried by some savages. Markhoor horns are of two types, both heavier than those that grow on Persian goats' heads, wild or tame. Both twist, the first openly like a pair of giant corkscrews, the other tightly like two great auger bits.

19

We began our list of wild goats with the Tur. We shall end it with the Tahr. The Tur is sheeplike. The Tahr is antelopelike. Unlike other goats, but like antelopes, the Tahr doe has four teats, instead of only two, for feeding her one Tahr baby. The beardless bucks go into the annual Tahr tournament wearing their short head-hugging horns like helmets. They battle and breed in the high Himalaya Mountains of Asia.

Also in the mountains of Asia are various goatlike antelopes, the little timid beardless Gorals, the homely shy Serows, and the rare Tibetan Takins. In America we have the shaggy snow-white antelopes known as Rocky Mountain goats. And living in Europe are the little Alpine antelopes called Chamois.

TAHR—an antelope-like goat

GORAL TAKIN CHAMOIS

SEROW

ROCKY MOUNTAIN "GOAT"

GOAT-LIKE ANTELOPES

Some of the handsomest true antelopes have horns resembling those of goats: the Koodoo of Africa and the Black Buck of India wear splendid spiraling weapons reminding us of Markhoors. Roan and Sable antelopes of the African plains carry great curving crests like the mountain-loving Ibex.

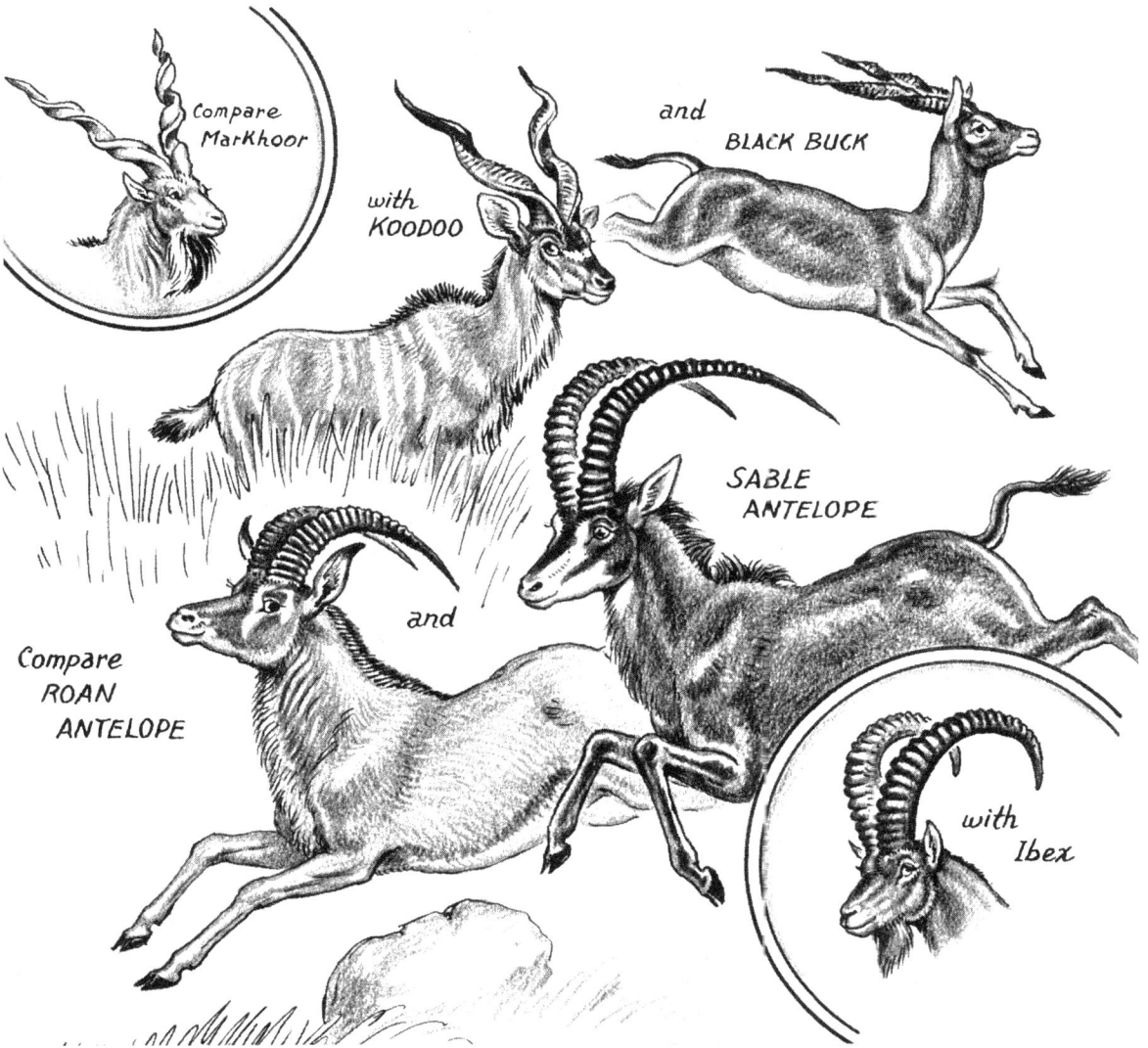

Compare
MarKhoor

and
BLACK BUCK

with
KOODOO

SABLE
ANTELOPE

Compare
ROAN
ANTELOPE

and

with
Ibex

But now let's have a look at tame goats. As with wild goats, the does of tame breeds are smaller than the bucks. So are their horns if they have them. But whereas wild does seldom if ever wear beards, a small natty beard is stylish among some tame varieties. One thing all tame goats share with wild ones—they never feel dizzy. Mountains are a goat's true dwelling place. With a perfect sense of balance, and aided by their sharp-edged hoofs, they walk undaunted on the steepest slopes where few other animals of such a size could find a foothold. They can stand on a pinnacle, with barely room to place all four feet, and jump from there to a narrow ledge with no concern for the awful chasm half a mile straight down below them.

No wonder our tame goats still happily climb to the housetops if given a chance. No wonder they do tightrope tricks in circuses. No wonder they do so well on

GOATS SPREAD HOOFS & GRIP THINGS FIRMLY. CAN THEIR (TOES)

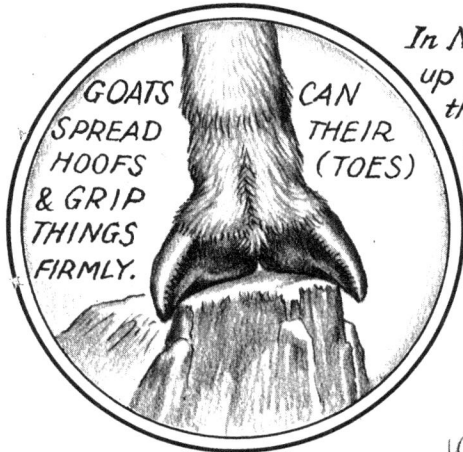

In Norway goats are sent up on sod roofs to "mow" the fresh grass sprouting there.

Clown and Acrobat

shipboard. Can an animal that never gets dizzy feel seasick? Many a vessel, especially in the Navy, has had a wether buck for mascot. On passenger vessels many a nanny has made a successful voyage to England from India or North Africa, nibbling at a bale of hay. She went

23

along to provide fresh milk in the days when ships had no refrigerators. These does remained in England and were bred to English bucks. This was the beginning of the Anglo-Nubian breed now called simply Nubian.

A way, way back, thousands of years before there were any ships, when all men still were savages, they weren't interested in goat's milk at all. They hunted goats with stone-pointed spears only for their meat and hides and useful horns. But then, about eight thousand years ago, hunters must have started carrying little orphaned kids home to their caves or huts, alive. The wild men, raising the little goats as pets, became the world's first herdsmen, and its first milk drinkers too, no doubt.

The first herdsmen gained a more reliable, steadier source of food and clothing than they'd ever had through hunting only. And so with the help of their goats and tamed wild sheep, they took the first big step in becoming civilized. Later on, when people had formed great nations, artists put goats in many of their wall paintings, tapestries, and sculptures.

In such works, still preserved through thousands of years, we see that some of their goats were Ibexes and others the Persian kind. We see that goats were valued by the ancient Hindus, Babylonians, Assyrians, Egyptians, Greeks, and Romans. Goats must have been among the flocks and herds of the ancient Hebrews in their great exodus from Egypt. And surely inquisitive goats tried to crowd in with the shepherds and sheep to see the wondrous sight in that stable at Bethlehem on the first of all Christmases.

Old-time Hindu tile design

Bas-relief from ancient Egyptian tomb

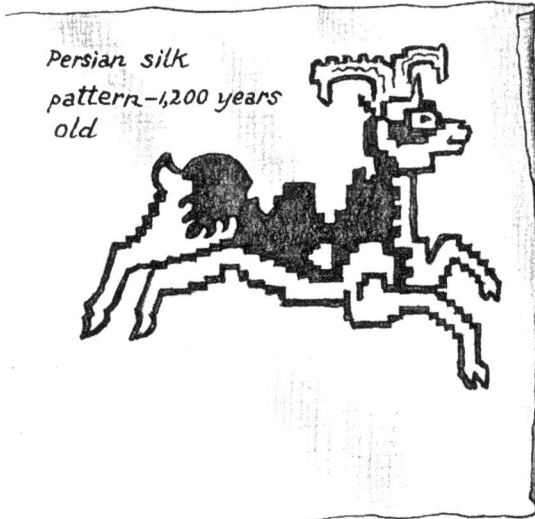

Persian silk pattern—1,200 years old

Prow of ancient Egyptian toy boat

Though civilization kept developing, most people still believed in many gods and spirits, some great and powerful, some merely fairies. They had a god of the winds, of the sea, of the sun and the moon, a spirit for every hill, a wood nymph for every tree, water nymphs for ponds and streams, and so on. People thought that gods and spirits could take the shape of animals to do them harm or favors, and had quite a list of animal gods and half-animal gods. The ancient Egyptians living in the city of Mendes had a goat god by the name of Ba. One of the many gods of the ancient Greeks was the half-goat god called Pan.

Except for the Devil's tail,

it was easy to be mistaken. WSB

Pan was the special god of flocks and shepherds, including goatherds. He had a human head and body and arms, but a goat's hind legs. There were horns on his head, his ears were pointed, and he had a beard and a goat's short tail. Along with certain godly powers, he had the mischievous, playful disposition of a goat. He liked to jump out and startle lonely travelers in the wilder places. And no doubt he was blamed for spoiling many a holiday in the country, turning a picnic into a pan-ic. That is how the word "panic" came to be. It means sudden, unreasonable fright, with perhaps a stampede such as flocks of goats make at times for no apparent cause. Pan is even said to have gained several military victories for the Greeks by panicking and scaring off attacking armies. Maybe the enemy thought he was "the Devil himself," also often pictured with horns, goatee, and cloven hoofs.

27

Pan may have done the ancient shepherds and goat-herds a very good turn too, for he was supposed to have invented the kind of pipes they used to play. Goats, like cows and many other creatures, enjoy music very much, so perhaps the ancient goats were happier and more con-tented and gave more milk because of the goatherd's music on the Panpipes. Why not? Goats give more milk today with a radio playing not too loudly in the milking room.

For centuries different tribes of people kept moving into Europe from the East. Some may have caught and tamed the wild goats already there when they arrived. But many must have brought their own flocks of the

28

Persian kind. The European hills and mountains proved ideal for these, and, as time passed, many fine types of goats developed, such as the several Alpines, the Toggenburgs, and the Saanens there today.

And long, long before those Nubian nannies sailed from Africa to England, somebody had brought goats of the Persian style to the British Isles. Probably the Phoenicians took them there about three thousand years ago. The Phoenicians were the greatest ship builders, navigators, and merchants of their times. Among other voyages, far for those days, they sailed to Cornwall to trade for tin. Goats may have been one of the things they exchanged for that metal with the ancient Celtic people living there.

Temporary "dimness of sight" while being "cured" of dizziness.

The Celtic priests, the doctors of their day, dealt as much in magic and spirits as in medicine. And they developed some strange prescriptions for healing the sick by the use of goats or portions of goats. For example, goats were thought useful in curing "falling sickness," perhaps because they never get dizzy. First the priest would apply a dog's gall to the patient's head. Then he would burn a goat's horn, blowing all the foul-smelling, suffocating smoke at the poor fellow, who would rise up and never have dizzy spells again, so it was claimed.

To cure dimming sight, the priest, boiling a goat's liver, had the patient keep his eyes in the rising steam. To cure deafness, he dropped melted and salted goat's fat into a person's ears. Burning a goat's hair was supposed to drive serpents away (and a lot of other things no doubt!). Ashes from its charred hip bones were considered a good tooth powder. Its blood was used as a cure for poison, for the bites of "creeping worms" and scorpions, for indigestion, and other troubles in people's insides. It was said to be fine for making a person perspire.

30

From early times, in various places, goats have been found "useful" in another, very mystical way—as scapegoats. Pagan people have always thought that every so often they must offer something of great value to terrible gods that otherwise would destroy them all. Goats and cattle might be all the wealth such people had, so they would trade a goat's precious life to preserve their own. Or on a certain day each year, all the sins of a whole community would be confessed over a goat; and then the animal would be taken out and lost in the wilderness. It was supposed to carry away all the confessed sins with it. Thus the wrongdoers 'scaped or escaped punishment.

Even now, in undeveloped areas, to end an epidemic or plague, all wrongs done in a stricken village are confessed over some poor goat and it is banished, the people praying that the plague will go with it. Or where only one person is ill, a goat may be brought into the sick room and then driven away with the same foolish hope. Such old-time notions die very slowly. In our own up-to-date America, a goat may be found living in many a horse stable. It is not there as a playful mascot. Persons who own and work with horses are sometimes very superstitious. They still feel that the goat may catch and wipe out any disease that otherwise would harm the horses. For the same magical purpose, goats may be placed in pastures with the cows—scapegoats one and all.

But here and now, for the most part, goat magic that can heal the sick and keep the healthy well is looked for only in the goat's milk. When the first settlers came from Europe to America, they brought a good many milk goats. They knew that these smaller browsing animals would forage more ably than cows along the fringes of the unbroken wilderness. But as the land was cleared, forests giving way to grass and grain, more and more cows were brought from overseas. The once-so-useful goats were

LIKE "MUTT" DOGS, SCRUB GOATS, THOUGH COMICAL IN APPEARANCE, OFTEN HAVE SUPERIOR BRAINS.

thereafter much neglected. They rapidly became a run-down race of scrubs, just "common American" goats.

For generations few people gave them a thought except as the butt of jokes. Then at last, early in this century, certain foresighted citizens noted that farms were growing smaller and city suburbs larger. They saw that in many of these small properties goats once again would be more suitable than cows. They imported purebred bucks from Europe and crossed them with common American does, and with the does' daughters, granddaughters, and so on. Steadily the scrub qualities were all bred out until today there are many big dairy herds and smaller flocks of purebred Nubians, Saanens, Alpines, and Toggenburgs in the country.

Milk-giving records of these four main breeds are all so good that one may as well choose which kind to keep from their looks. If you were planning to be a goat dairy-

33

NUBIANS

If scrubs are funny —

so are galloping Nubians. They seem to fly with their ears !

man, which would you prefer? Judging by their pictures, would you fancy Nubians? They are long-legged and short-haired. The commonest color of the slick coat is a deep ruddy-brown shading to black. Yet some are splashed all over with a pattern of big white polka dots besides. But their most notable features are the high arching profile of the face and the long, broad pendant ears, almost like a bloodhound's. The beard is small. So are the horns, if any.

In every herd of wild goats there is always the possibility that a kid or two will fail to develop any horns. Now, wild goats with no horns have little chance of growing up and having kids as hornless as themselves. They are sure to lose their lives before that happens in battles with well-horned goats or with beasts of prey. Only the well-horned

34

members live, and almost all the kids will grow good horns just like their parents.

But in a tame herd, protected by fences, goatherds, and sheep dogs, the hornless kids get along very well. Indeed the dairyman-breeder treats them with especial care. And he breeds them together so that they may produce more hornless goats. This has been going on for so many generations that today (at least in Europe and America) the majority of milk goats have no horns.

Of course it is more normal for goats to have horns, and so there are always quite a few tame goats that sprout the ancient armor. This does not please the dairyman, for even if only in play, a prize-winning doe might be so injured by another goat's horns that she'd never give another drop of milk. The fewer such accidents, the better for his herd and the business. He can operate to prevent a kid's horns from ever growing, or if any of his grown goats already have horns, they can be removed in several ways.

Because this is necessary, a dairyman is inclined to think a hornless goat is prettier than one with horns. But this has more to do with success in business than with beauty. To an artist who lives by other work, it seems a pity that horns must be prevented or removed. To him, a goat complete with splendid, graceful headgear is far more beautiful than one without. At least one famous goat breeder felt the same way. He preferred possible butting accidents to the certain pain and damaged feelings of a dehorned animal.

Returning to your own preferences, instead of the dark short-haired Nubians, how would you like to have some big blond Saanens? Their rough coats are all white or cream color, like milk itself. Seen against the bright green of a lush spring pasture, Saanens are as charming a sight as a flock of snow-white sheep.

SAANENS

Or maybe you would favor Alpines. They range in color from white to black through various shades of tan

36

ALPINES

SHAKO

CHIN STRAP

and gray. Quite commonly they are a light tone on the forward half and a dark shade aft. And often also, the costume has a very military look with stripes on all four legs, a stripe all along the spine, and one or two down each cheek like the straps of a gay shako.

TOGGENBURGS

But of course, why not Toggenburgs? Tried and true, they were the first milk goats to be cultivated widely in

37

this country. All wear the same pattern of brown and white, though some are like chocolate and maple ice cream while others are a lighter cocoa and vanilla. The white is mostly on the face and ears, the belly and rump, and the inner side of the legs, rather like a deer.

So there you are. Choose according to whichever pleases your own eyes most. In ten months' time a merely "fair milker" of any one of the four breeds will give more than one thousand pounds of the highest quality milk. The very best milker will give four times that much. Goat's milk is so much like the milk of human mothers that little babies who cannot take cow's milk, or any formula, do well on it. It aids ailing people too, people recovering from tuberculosis, people with eczema, weak digestion, stomach ulcers, and so on.

It is a delightful, wholesome food for healthy persons as well. Why, even Zeus, greatest of all the gods of the ancient Greeks, was supposedly nursed and mothered by a goat named Amalthaea. Whether they really believed that or not, it shows how well the Greeks knew the worth of a goat's good milk long, long ago. Goat's milk is used for various animals too. Orphaned calves and lambs that otherwise would have died have been nursed by nanny goats.

In warm lands, many a nanny not only nurses a child while his mother is away at work. The goat adopts the baby and watches over him. She will leave her browsing and come a-running when he cries.

Have you ever wondered about the mystery of milk? Not how it got into the bottle in your refrigerator, all the way from a cow on a farm to the bottling works to the store where your mother bought it. That is not the mystery of milk. The mystery is: how did the milk come to be in the cow in the first place? By what magic in her body was the water she drank and the grass she ate changed into milk, the "almost perfect food"? How does the same thing happen to weeds eaten by a goat, to a lioness's meal of zebra, to anything any mammal mother dines upon?

Scientists have been working on this mystery for years. By now they know most of *what* happens inside bossy or nanny as milk is made, but still they are stumped concerning much of the *how*. They know that a meal of clover, cropped and swallowed with almost no chewing, goes to fill the first of a goat's four stomachs. There it is formed into neat balls called cuds, which rise into the goat's

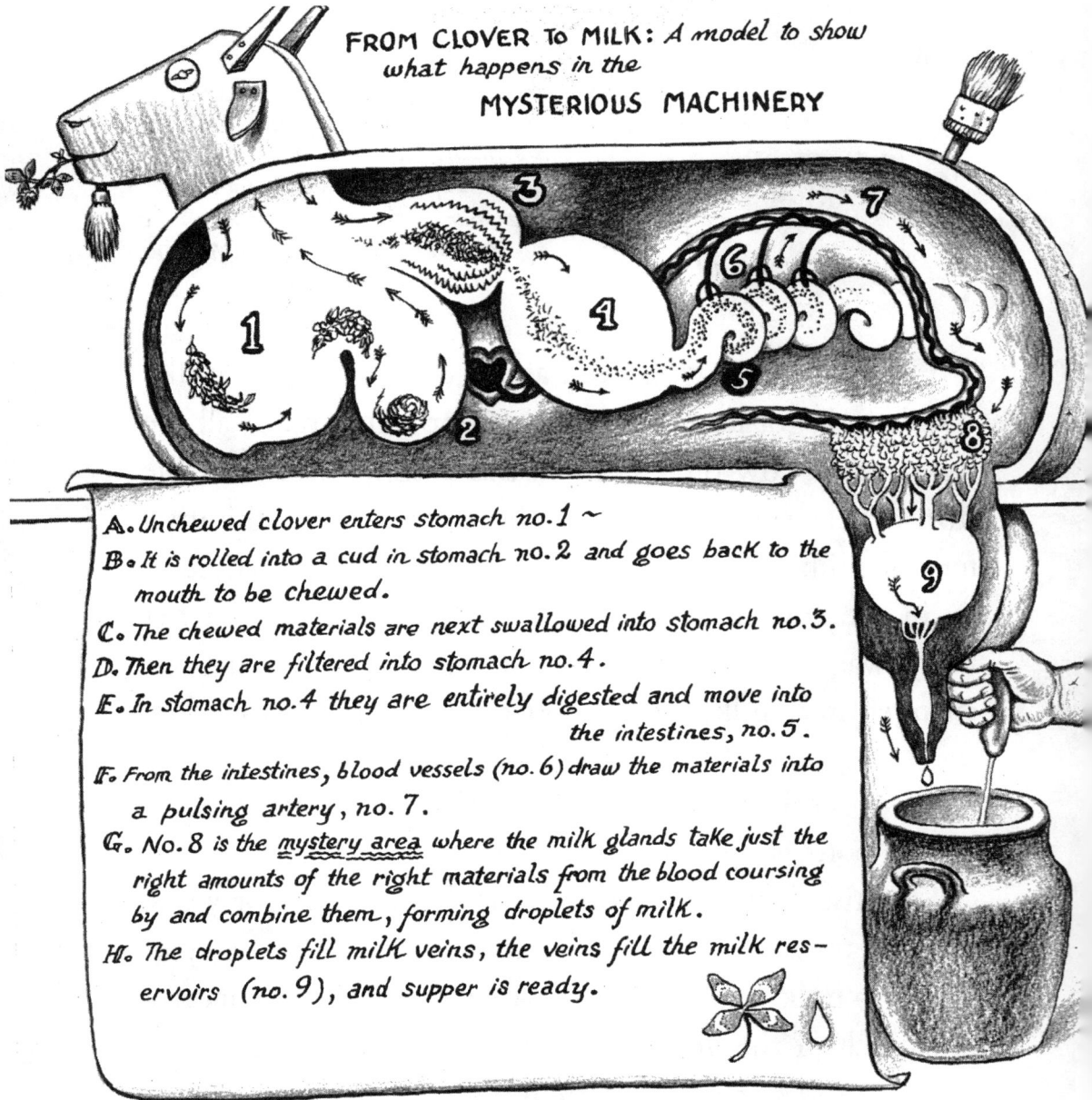

FROM CLOVER TO MILK: *A model to show*
what happens in the
MYSTERIOUS MACHINERY

A. Unchewed clover enters stomach no. 1 ~
B. It is rolled into a cud in stomach no. 2 and goes back to the
  mouth to be chewed.
C. The chewed materials are next swallowed into stomach no. 3.
D. Then they are filtered into stomach no. 4.
E. In stomach no. 4 they are entirely digested and move into
                              the intestines, no. 5.
F. From the intestines, blood vessels (no. 6) draw the materials into
  a pulsing artery, no. 7.
G. No. 8 is the mystery area where the milk glands take just the
  right amounts of the right materials from the blood coursing
  by and combine them, forming droplets of milk.
H. The droplets fill milk veins, the veins fill the milk res-
  ervoirs (no. 9), and supper is ready.

mouth one by one to be well chewed and swallowed a
second time. As this chewed-up clover passes through

40

stomachs two, three, and four, the goat's digestion breaks it down into all the separate nourishing substances of which clover is made.

The scientists know how these substances are soaked up by the goat's circulation and carried to all parts of her body, feeding every bit of it including her bag or udder. But in the bag something magical occurs, turning a portion of those clover materials into milk to feed her babies. How this happens is not yet fully known.

As the miracle takes place, thousands of minute milk droplets form, each with just the right amounts of everything a perfect food should have. They trickle through tiny milk veins into veins a little larger, joining others lower down and larger still, like rivulets rising from springs of water in a lush, sweet meadow.

And just as the rivulets join, forming streams that flow at last into a lake, the veins of milk finally reach a big reservoir in each half of the bag. Here much of the milk collects. From these big reservoirs many little veins carry some of the milk still farther down to a smaller reservoir in each teat. And so when the kids suck or a hand squeezes it, one finger after another, the milk flows into mouth or pail immediately.

The bag of a wild goat never grows very large. It holds just enough milk for her kids. When they are weaned and stop suckling, the milk stops forming. Then the bag shrivels up till the next time she has babies. For a wild nanny that climbs among rocks, browses in thorny brush, and sometimes fights or runs for her life, a big bag would be very bad. It would soon get all cut up, be very much in the way, and slow her down so much that she might lose her life.

But the dairyman keeps milking his tame, protected goats for many months after the kids are weaned. Their bags grow bigger, giving larger amounts of milk each time new kids are born. Thus he has enough for the kids, some for his own family, and plenty of milk to sell.

Does a goat dairyman's or a breeder's work seem like a good and pleasant sort of business? It is, but making a success of it takes a lot of doing. A person has to like animals well enough to work with them every day seven days a week, and sometimes all hours of the night, year in and year out. He needs to know every goat in his herd for its

It is not uncommon for kids to be born at 2 A.M. regardless of the weather.

special traits, likes and dislikes, just as he knows people. Each one wants to be treated like a special friend, to have a special share of his affection, to feel the caress of his hand, and to hear kindness in his voice.

If he can give them that, the work will be fun and they will give him all the milk they can with all their hearts. If he doesn't really like them, nor they him, they can "refuse to let down their milk." Perhaps they don't actually hold it back on purpose, but the milk really doesn't flow as freely or as plentifully.

Methods vary from goat farm to goat farm. But let us suppose that at a certain dairy eight goats are let into the milking room at a time. For each doe there is a trough full of the day's big treat, beet and orange pulp mixed into a sweet mash with warm water and molasses. They all eat

this eagerly while being milked. Now goats are not only animals but individuals too. Each has its own special feelings and ideas. Most of them are naturally affectionate, but some much more than others. Such a one, although equally hungry, may so dote on her master that she won't begin feasting with the rest. Instead, she will stand in the middle of the room calling loudly and only go to her food trough when he comes in. To her, he is the most important person in the world.

Perhaps you will be that kind of goat dairyman some day. There are many things to know and do for and with a herd of twenty goats or more. You will have to know how to feed them so that their kids will be born healthy and grow fast and strong, and how to feed and breed them to get the most milk possible through the year. You will work as a barber, clipping and brushing their coats, and as a manicurist once a month or whenever their hoofs

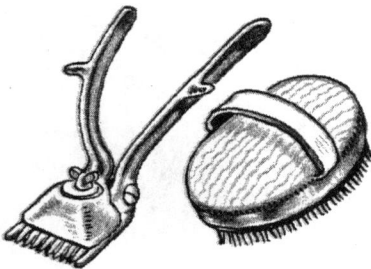

PRUNING SHEARS ARE FINE FOR TRIMMING HOOFS. TRIM THE HORNY RIMS JUST EVEN WITH THE SOFT TOES INSIDE.

KEEP A NANNY NEATLY CLIPPED ABOUT HER BAG. GIVE YOUR GOATS A DAILY BRUSHING. TO THEM IT IS LIKE A BATH.

CLIMBING ABOUT AND RUNNING ON ROCKS, WILD GOATS KEEP THEIR OWN HOOFS WORN DOWN LEVEL AS FAST AS THEY GROW.

If the corner braces are inside the fence (the usual place), goats will walk up them and get out. Some goats may try to dig out. All bark will be nibbled from the posts in short order.

need trimming, and as a doctor for them at birthing time or in case of accidents or sickness. Of course, if the trouble is very bad, you can always call the real animal doctor, the veterinarian.

The dairyman must make and maintain strong fences, build and keep clean stables, milking room, milk-cooling room, and feed-storage room, sterilize his equipment, bottle the milk, find new customers, and deliver the goods every day no matter what the weather. He has to know and obey all the laws for sanitary handling of the milk in order to pass inspections for the cleanliness of his buildings and the health of his animals. Fresh clean water must be available to them at all times, and it must be warmed in winter. There must be blocks of salt, mixed with other minerals, for them to lick as often as they please. Besides

45

all these items and many others, there is, as with any other business, the cost of getting it started and keeping it going —first of all the equipment and then the ever repeated purchase of hay and feed and other supplies. So, a complete goat-milk business is something to plan very carefully for quite a while before actually going into it.

In the meantime, there is a fine way to learn a lot about it and gain both pleasure and profit too. Just for fun, let's pretend for a moment that your dad has a constant nervous tummyache. Or maybe Aunt Mirabel is very frail and Grandpa can't digest cow's milk at all. Or perhaps cow's milk gives baby brother the hives. Then, as you already know, goat's milk would be just the thing for your family. It is delicious. All healthy people can enjoy it. And it would mend your family's ailments in short order.

But possibly there is no goat's milk to be had as yet in

your part of the country. Then why not send to the nearest good goat breeder for two nice does and have your own supply right there at home?

First, you would find out if local rules permit the keeping of livestock. They may allow nanny goats but no billies. A billy goat might "set the neighbors by the ears," or noses. But you wouldn't want a buck for only two does anyway. When the time came for them to be bred again, they could go back to the breeder's farm for a little visit. Then five months later there would be newborn kids and increased amounts of milk at your house.

If you could get one doe that would have her kids between Christmas and Washington's birthday, and the other between Decoration Day and the Fourth of July, you would have plenty of milk for all your needs all through the year. Whatever they cost would soon be canceled out, for drinking and cooking with your own goats' milk would be far less expensive than using store milk. Of course, though you'd spend no more money for milk, you would spend some of your time. But you shouldn't count that since you'd be working for yourself.

Naturally it would be well to learn all you could about milk goats before getting any: how to buy wisely, how to house, feed, and milk them, when to breed them, how to cure and (more important) to prevent sickness, how to raise the kids, and how to ready them to try for prizes in the livestock shows. Raising a couple of goats from kidhood to milk-producing does would be a good project for a 4-H Club member or for a Future Farmer of America.

A few pages farther on we shall consider some dos and don'ts about the proper housing and feeding of goats. This will help you some. But there are other books, special guides for good dairying, as well as bulletins and livestock magazines, that are full of information about goats and how to handle them. Some of these might be in your local public library, or you could borrow them from your state library in the state capital city. Your State Extension Service at your state university would send you booklets and bulletins free of charge. They could give you the names of goat societies and goat magazines and journals. Your State Dairy Commissioner, your County Agent, your veterinary or animal doctor, and your local feed dealer could all help you. So could the Bureau of Animal Industry, U.S. Department of Agriculture, Washington, D.C.

In the goat magazines you'll see advertisements of all the best goat breeders, one or two surprisingly nearby perhaps, close enough to visit and see their animals and get acquainted. The magazines also carry advertising for all kinds of equipment to aid the goat keeper. And one of the editors would answer a letter of inquiry if you were polite enough to enclose a stamp.

It is better to begin with two does rather than one. Alone among strangers, a goat may be so homesick that she will give less milk than she did on the farm. Caring for two is far less than twice the work; but you get over twice as much milk from two happy goats as you do from one that is sad. And just as it is best to begin with at least two goats, it is best for at least two people to have their care—two brothers, a brother and sister, two sisters possibly.

Perhaps Dad would pay the young goat keepers the money otherwise spent buying milk for everybody week after week at the store. He might like to join his children in the work as a fine, paying, healthy hobby. A schedule could be planned so that morning and evening, before and after school, someone would be on hand without fail to do the chores, the milking and feeding and tidying up. These things can't be put off or skipped even once. Your animals have to depend on you. And anyway, like other intelligent pets, goats very quickly become members of the family to whom you wouldn't begrudge the time or anything else in reason.

However, maybe you don't really want or need a homemade milk supply. Yet you might wish to have a good pet wether goat or two, like the one described in earlier pages. A good pet wether can be of any breed or mixture. He could very well be an Angora with long spiraling horns and long silky fleece.

ANGORAS

Angora goats are raised for their wool and meat, but not for dairy uses. They bear but one kid at a time and, like wild goats, have only just enough milk to fill its little stomach for a few short weeks. Most big Angora flocks are in our western states. There they not only supply the wool for weaving mohair cloth, but they are also used to browse all the brush and weeds from land where the forest has been cut and crops are to be planted. They also keep firebreaks and railroad rights-of-way clear in the same manner.

And they make beautiful pets if you can keep them out of burrs and brambles. One Angora or a team hitched to a neat little cart is a pretty sight and can provide a lot of fun. So can wether bucks of any kind. They can work for you too. Fully grown, one such powerful pet can pull

a small dump wagon in which you might deliver goat garden fertilizer (the best in the world) to people who would buy it for their shrubs and flowers. Or perhaps you'd prefer to use the fertilizer on your own garden to grow fine vegetables. Then your pet could drag the garden cultivator and later help deliver vegetables to housewives. No doubt you'd think up other things to do with your strong and willing friend.

Watch him when you halt. He might try to nibble your produce.

It is not wise to buy a baby goat less than three months old. Younger than that, a kid is not ready to leave its home and mother, its playmates, and the care of an experienced dairyman or breeder. But training to harness

should begin as soon as your young wether is used to his new home and looks upon you as good company and the supplier of all his needs. He will be glad to oblige, once he understands what you want him to do. But at first, the bit in his mouth and your "G'dupp!" and "Whoa!" as you try to lead and halt him, won't make sense. Neither will the cart you attach to his body, once he learns to obey your voice and the reins. To him it will only be a large object strangely trailing behind him everywhere he goes. But finally he will "get the hang of it all" if you are patient. You must humor him. And who knows how much he will be humoring you? A switch will only make him stubborn as a mule. Train him on a half-empty stomach. Give him some hay and grain before the work-out but no greenery. If allowed to graze and browse first, he'll bulge too much. His harness straps will be too tight, and he'll be blown in no time.

53

Here are a few hints for having a healthy, happy pet wether buck. Before buying him, have his living quarters ready. There should be room enough inside for him to move about a bit, sleep out of drafts, and be fed and watered. You will want a partitioned-off place to store his feed and baled clover or alfalfa hay and bedding material. Keep these needs in mind, plus the size he may possibly be when fully grown. Goat kids and human "kids" average about the same weight at birth. Some of each grow up to weigh two hundred pounds and over. Such a buck will be about five feet long and stand some forty

A three-sided shelter facing south is good – one in the pasture for sudden wet weather, and one against the barn where goats can gather at milking time.

HAY

WHITE WASHED WALLS & CEILING

DOOR TO HAY & FEED ROOM. KEEP IT LOCKED.

WATER

GRAIN & FEED

SALT

FERTILIZER TROUGH

PUT HAY OVERHEAD. GOAT WILL WASTE LESS AND KEEP HAPPILY BUSY — GETTING IT.

inches high at the shoulders. So, plan his home accordingly, just in case.

The dwelling may be a tidy shed built especially for your animal friend or a converted henhouse or other small building. Or it can be an area set aside in the barn if you have one. It should be well lighted by a window he cannot reach with hoof or horn. It must be well ventilated but not drafty. It must be dry. Goats are tough in many ways. They don't get tuberculosis as cows do. But they aren't proof against pneumonia. They can stand cold, but dampness is a danger to them always, and they know it. They will stampede from a light midsummer shower as

55

Goats generally lie down like cows (forelegs first),

but often like horses, (rear legs first),

and sometimes with all legs folding almost at the same time.

2 days old

Even a tired baby can make its own bed, and does so.

though their very lives are in peril. No cat ever detested wet feet more than goats do.

So, the floor of your wether's house must be well drained and above the level of the ground around it. Concrete is easiest to keep clean, but it is cold and hard to stand or lie on. Goat men disagree on flooring and bedding. Some say concrete is fine, or slatted wooden platforms, and that no bedding is necessary. Others claim

On high mountain ledges, one goat will leap over another to reach a place to lie.

that whatever the flooring, there should be bedding of some sort. But why not settle the argument by observing the goats themselves. Their vote would be unanimous in favor of bedding, for indoors or out, before lying down, every goat, from a day-old kid to a bearded ancient, will rake together with a front foot whatever straw or rubble is handy for a makeshift mattress. Wild goats in their craggy mountains undoubtedly do the same, and choose to lie, if not on the "softest" rocks, at least on the mossiest.

Therefore, whether the floor is concrete or wood, give your pet a really restful place to lie at night and in bad weather. Peanut shells or other cheap litter can be had from your feed dealer. Each day you can remove whatever has gotten moist and mix it with the manure to molder and become a part of the world's finest fertilizer.

Goats are lovers of liberty. Only wild ones have it completely. The most freedom a tame goat gets is all within a well-fenced pasture or from whatever wandering a watchful herdboy and his dog allow. For one or two goats you need at least one acre of browsing and grazing land. You don't have that much? Only a few building lots? In that case you could stake your pet. This should be in a fresh place every day. If your own grass area gives out, some of the neighbors might let you stake your goat on theirs now and then, for goats are the best of all natural lawn mowers; better than cows that crop the grass unevenly, and better than sheep that cut it too close to the ground.

"Yes ma'm. He'll keep at it all day long except for a little time for chewing his cud."

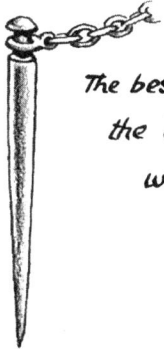

The best stake has a swivel so that the chain can turn and not wind up around it.

Drive it in just level with the ground.

Use a chain for staking, not a rope. A rope can tangle and strangle. It kinks and snags on things. And if it doesn't come untied, any self-respecting goat will finally work it loose himself. Use only a dozen feet of the chain at first so that your beginner won't hurt himself when brought up short. Lengthen it as he learns its limits. He will very quickly. He may also prove his intelligence by learning to keep his chain free although staked among a lot of tree stumps or other objects.

A very tough liberty-lover may even discover how to <u>use</u> a stump to break a <u>rope</u> by repeated rushes.

Once more

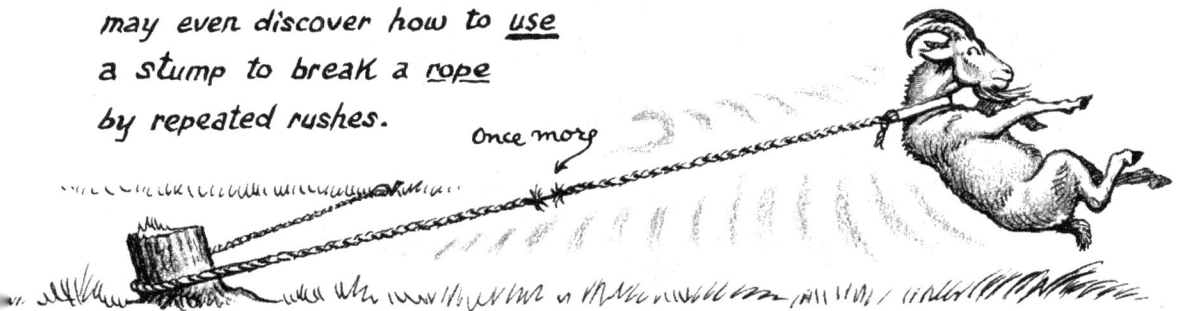

If dangerous or merely playful dogs, or thoughtless people, annoy your goat, a neat corral would be better

He will spend much time on his private "hill-top" watching the world beyond, and less time trying to butt his way through the seven-foot-high stockade.

PACKING BOXES

An empty nail keg will amuse him for hours.

than a chain. Perhaps it would be anyway, for wherever he is outdoors, he will need not only sun and shade and a shelter he can reach in case of rain; he will also need some chance to play and prance unhampered by a chain, something to climb and jump from, and something to butt. A corral can be furnished as in the picture, a sort of goat's outdoor gymnasium.

In a corral every blade and leaf of greenery is soon eaten. You will have to supply all the fodder in summer as in winter. Besides hay, grain, and prepared goat foods from the feed man, you can treat your goat, from the kitchen, to clean potato, turnip, and fruit peelings, to the outer leaves of cabbage, and to stale bread too, unless Mother makes bread puddings.

Carrots, beets, and all such vegetables are fine, but never offer wilted stuff or anything that has been frozen. A sensible goat may decline the latter, but if he eats it, he's almost certain to be sick. Give him garden weeds and the cut lawn grass and clippings from the shrubbery—not from laurel or rhododendrons though. They are death to sheep and goats. Privet berries are poison also. You can get lists of plants that are poisonous to livestock from your State and Federal Departments of Agriculture.

In the agricultural bulletins and pamphlets you will find the proper amount of various foods to give a goat. It differs of course from goat to goat. But you will soon discover how much each of yours cleans up at a feeding and adjust the amounts to suit them and waste very little. As a trial winter ration, you might start with two pounds of alfalfa or clover hay, a pound of root vegetables or beet and citrus pulp, and one-and-a-half pounds of grain, half offered in the morning and half in the late afternoon. The grain might be a mixture of corn, oats, bran, and linseed meal.

In summer, if you have pasture, the amounts of hay and grain can gradually be reduced as the summer comes on and the succulent roots omitted while there is plenty of grass. A milking nanny will use about five hundred pounds of hay and four hundred and fifty pounds of grain

a year, if there is some pasture for six months of that time. All in one heap this would look like a lot, but it is only one eighth of what a cow must have. A buck on good pasture needs no grain in summertime at all.

Neither does a wether. And as for keeping such a jolly pet, it might be twice the fun to have a team instead of one wether alone, since goats are so fond of company. It almost seems that goats are commonly born as twins simply to be sure they'll have a pal to play and butt with. In fact, one almost wonders if the twins play at butting even before they are born, since so much activity can be felt by touching the mother goat's side. Very soon after birth, even with their fur not yet quite dry, baby goats will try to climb up on things and in a week are playing King of the Castle on some hummock.

It's the same on the level with no hill to win or lose.

63

They butt. And they butt. Suddenly they race off together. They leap and twist in the air. Sometimes they even turn somersaults. Then they butt some more. Such are the capers of capricious *capri*. (Capri is Latin, the ancient Roman word for goats.)

Have you noticed in the pictures that most of the goats are drawn with the corners of their mouths turned up a little? That is how they really are. Much of the time goats seem to be smiling. It suits their natures perfectly too, for goats are naturally gay. And if you care for your nannies or your pet wethers faithfully and well, you will have their affectionate and lively company for a dozen years to come, or even more.

BUCK = A he-goat.
BUCK = To butt.
Bump = A jolting knock.
Bump = To crash into—
Bunt = To push with the horns.
Butt = To strike with the head.
Butt = The END of a thing.
Butter = One who butts

WSB

www.ingramcontent.com/pod-product-compliance
Lightning Source LLC
Chambersburg PA
CBHW081240090426

42738CB00016B/3362